ZOOM IN!

For more than 50 years, Guinness World Records has documented the world's ultimate record-breakers in every field imaginable. Today, the records in their archives number more than 40,000.

In this collection, we'll zoom in on 25 record-holders in the mechanical world. We'll power up with natural energy and go places inside souped-up vehicles. Next, we'll take apart the gears and gadgets of our favorite entertainment devices. Then we'll run, jump, and feel with the newest robots whirring into our homes today. Finally, we'll go on a wild ride with five superstar record-holders that are the "Best of the Best!" Go up close and get revved up — Guinness World Records style!

A Record-Breaking History

The idea for Guinness World Records grew out of a question. In 1951, Sir Hugh Beaver, the managing director of the Guinness Brewery, wanted to know which was the fastest game bird in Europe — the golden plover or the grouse? Some people argued that it was the grouse. Others claimed it was the plover. A book to settle the debate did not exist until Sir Hugh discovered the knowledgeable twin brothers Norris and Ross McWhirter, who lived in London.

Like their father and grandfather, the McWhirter twins loved information. They were kids just like you when they started clipping interesting facts from newspapers and memorizing important dates in world history. As well as learning the names of every river, mountain range, and nation's capital, they knew the record for pole squatting (196 days in 1954), which language had only one irregular verb (Turkish), and that the grouse — flying at a timed speed of 43.5 miles per hour — is faster than the golden plover at 40.4 miles per hour.

Norris and Ross served in the Royal Navy during World War II, graduated from college, and launched their own fact-finding business called McWhirter Twins, Ltd. They were the perfect people to compile the book of records that Sir Hugh Beaver searched for yet could not find.

The first edition of *The Guinness Book of Records* was published on August 27, 1955, and since then has been published in 37 languages and more than 100 countries. In 2000, the book title changed to *Guinness World Records* and has set an incredible record of its own: Excluding non-copyrighted books such as the Bible and the Koran, *Guinness World Records* is the best-selling book of all time!

Today, the official Keeper of the Records keeps a careful eye on each Guinness World Record, compiling and verifying the greatest the world has to offer — from the fastest and the tallest to the slowest and the smallest, with everything in between.

Compiled by Celeste Lee and Ryan Herndon

For Guinness World Records: Laura Barrett Plunkett, Craig Glenday, Stuart Claxton, Michael Whitty, and Laura Jackson

SCHOLASTIC INC.
New York Toronto London Auckland Sydney
Mexico City New Delhi Hong Kong Buenos Aires

Attempting to break records or set new records can be dangerous. Appropriate advice should be taken first, and all record attempts are undertaken entirely at the participant's risk. In no circumstances will Guinness World Records Limited or Scholastic Inc. have any liability for death or injury suffered in any record attempts. Guinness World Records Limited has complete discretion over whether or not to include any particular records in the annual *Guinness World Records* book.

Guinness World Records Limited has a very thorough accreditation system for records verification. However, while every effort is made to ensure accuracy, Guinness World Records Limited cannot be held responsible for any errors contained in this work. Feedback from our readers on any point of accuracy is always welcomed.

ISBN-13: 978-0-439-89826-3
ISBN-10: 0-439-89826-9

Designed by Michelle Martinez Design, Inc.
Photo Research by Els Rijper, Sarah Parrish, Alan Gottlieb
Records from the Archives of Guinness World Records

12 11 10 9 8 7 6 5 4 3 2 1 6 7 8 9 10/0

Printed in the U.S.A.

First printing, December 2006

Visit Guinness World Records at www.guinnessworldrecords.com

Power Up

Who controls the world: humans or machines? People invented mechanical marvels to improve the quality of life. Machines carry massive loads, transmit electronic messages globally, and entertain and challenge our brains with visual mazes. But who, or what, runs the machines that save us time, get us there faster, and makes our lives fun and exciting? Let's discover what really powers up the ultimate record-breakers of the mechanical world.

Engineers transform raw power into usable energy.

Oil, coal, and fire are raw energy sources. Sun, wind, and water are naturally *renewable* energy sources. Engineers build mechanical masterpieces, such as dams and wind farms, to collect and convert nature's immense and fluid power into containable, usable energy for people. Hydropower comes from the force of falling water funneled through a dam. At the dam's peak, water flows through a fitted channel then falls into the reservoir below. Raising the dam's height increases the water's falling force and therefore the amount of energy collected.

RECORD 1

Largest Concrete Dam

During the Great Depression of the 1930s, President Theodore Roosevelt approved the construction of the **Largest Concrete Dam**. Building on the Columbia River, Washington, USA, began in 1933 with the Grand Coulee dam operational on March 22, 1941. Fully completed in 1942 at a cost of $56 million, this impressive structure is also a much-visited tourist attraction. It has a concrete volume of 285,760,000 cubic feet to a weight of 43.2 billion pounds at a crest length of 4,173 feet and 550 feet high — that's taller than the Great Pyramid of Giza and twice as high as the Statue of Liberty! Originally intended to irrigate the dry lands of the area, engineers soon discovered this goal was merely a drop in the bucket. Today, the Grand Coulee dam is the largest producer of hydroelectric power in the USA.

FACT:

Hydropower supplies 20% of the entire world's electricity.

Waterwheels and aqueducts harness hydropower. The Largest Waterwheel is Mohammadieh Noria at 131 feet in Hamah, Syria.

THE NEXT BIG THING

All energy sources have drawbacks. Dams and reservoirs displace wildlife and human communities, but hydropower's positives far outweigh any immediate negatives.

China's 1.3 billion citizens have enormous energy needs, with 80% currently supplied by burning coal. Hydropower would decrease environmental pollution. The 607-foot tall Three Gorges dam spanning the Yangtze River was finished in 2006, and its planned 26 generators should be operational in 2008. The dam is expected to provide 3% of China's electrical needs and become the new record-holder for the Largest Concrete Da

Feel the breeze on your face, and see wind power light up your home.

Fossil fuels, like coal and oil, create pollution and will eventually be gone. Wind power is a renewable energy straight from nature. Centuries ago, the wind's strength was harnessed for grinding corn and collecting river water to irrigate crops. Today, the wind turns giant blades that spin a shaft inside a small generator to create electricity.

FACT:

The Most Powerful Solar Tower Power Station is located in the Mojave Desert, California, USA. Solar II uses 1,800 curved heliostat mirrors to reflect sunlight onto a central heating lement and generate 10 egawatts of electricity.

RECORD 2

Largest Wind Farm

Since 1981, the **Largest Wind Farm** has generated 6 billion kWh (kilowatt-hour), enough electricity to power 800,000 houses for one year! This is the exact same electrical energy produced by generators burning coal, natural gas, or oil. The Pacific Gas and Electric Company's wind farm was built in response to the first oil crisis in the 1970s. There are 7,300 wind turbines spread across 54 square miles in Altamont Pass, California, USA. Altamont's turbines are shorter than the newer turbines available. The latest turbines are taller with larger blades and can even be installed offshore. Their superior height collects more wind because of less drag from the ground below, and higher altitudes provide wind day and night, so electricity can be produced around-the-clock. Some of the newest turbines are designed to collect wind *and* solar energy!

RECORD 3

Oldest Steam Engine

Waterwheels were a major energy provider until 1711. That's the year **steam power** began to push the Industrial Revolution forward. Economic growth could now occur in areas without depending upon a nearby river. However, it was inventor and engineer James Watt who realized how to make the engine run faster on less energy.

In 1779, Watt designed the Smethwick Engine to pump water back up into the canals in Smethwick, England. Built by the Birmingham Canal Company, this little engine had a 24-inch bore with a stroke of 8 feet and capably did its job for more than 100 years until 1891. The Birmingham Museum of Science and Industry in England regularly steams the original Smethwick Engine for the public's enjoyment. It is the **Oldest Steam Engine** still in working order.

Bubble over to Record 7 for more on steam-powered cars.

STEAM AWAY

A steam engine requires a boiler to first boil the water to produce steam. The hot water's kinetic energy will move a piston or blade in a continuous back-and-forth motion. This repetitive back-and-forth gradually turns wheels, gears, or pumps. This is an example of thermal energy at work. The advantage of the steam engine was its simplicity and mobility. All you needed was fuel to start a fire — wood, coal, or oil — and water in a boil-ready container before setting off on your trip . . . with lots of extra time to stop and reheat your engine every 10 minutes!

Before planes, trains, and automobiles, foot power got people where they were going.

Human Powered Vehicles (HPV) are odd and wacky machines on wings or wheels. All you really need are strong legs. In the mid-1800s, people rode recumbent or seated bicycles, pushing foot pedals and hand cranks. In 1933, a French cyclist named Francois Faure broke a 20-year-old cycling record on a recumbent bicycle called the *Velocar*. The governing body of bicyclists banned the recumbent bicycle design from competition, yet the public continued its fascination and development of these pedal-powered vehicles.

RECORD 4

Highest Speed on a Human Powered Vehicle

Power pedaler Sam Whittingham proves we can still get around fast using our own two feet. On October 5, 2002, the Canadian reached 81 miles per hour on his streamlined recumbent bicycle, achieving the **Highest Speed on a Human Powered Vehicle** at the World Human Powered Speed Challenge near Battle Mountain, Nevada, USA. He sped across flat ground, without the help of gravity or mechanical energy. His bike's builder, Georgi Georgiev, is redesigning the *Varna Diablo* to eliminate the tiniest wind resistance between the viewing bubble and shell. The new bike will completely enclose the rider. He must use a camera for navigation. Currently, Sam is training with a pillow atop his head to get comfortable with cocooning himself in the new bike design.

FACT:

A hydroplane is a flat-bottomed boat lightweight enough to skim across the water's surface.

HYDRO HYPNOSIS

Many different kinds of aquatic vehicles are used by international ferry and naval services. A hydrofoil is a boat with wing-like structures, called foils, mounted beneath the main body of the boat called the hull. As a hydrofoil moves, the foils cause the hull to actually lift up and out of the water. Getting the hull out of the water reduces drag, allowing the boat to go faster. In a hydrofoil, you only need to move the foils through the water, instead of the entire boat.

RECORD 5

Fastest Human Powered Speed on Water

Have you ever powered a pedal boat? They're fun, but their speed depends on your pedaling power. The *Decavitator* is a human-powered hydrofoil, developed at the Massachusetts Institute of Technology (MIT) in Boston, Massachusetts, USA. Designing the speedy hydrofoil took a great deal of trial and error. It was discovered that making the wings (also known as foils) smaller would allow the boat to go faster. Ultimately, the record-setting wing size was a mere 1.75 x 29.5 inches! Mark Drela, one of the boat's chief engineers, pedaled a record-setting speed of 18.5 knots (or 21.28 miles per hour) over a 328-foot course on the Charles River on October 27, 1991. At 31 feet per second, he achieved the **Fastest Human Powered Speed on Water**. Want to see the *Decavitator* in person? You can visit it at the Museum of Science in Boston.

Drag is the resistance experienced by a body moving through the water or air.

Going Places

Going from one spot to another is always a challenge. The solution lies in the choice of your vehicle. Our society has traveled a long way from wheels made out of stone to those that you can roll from the *inside*, and a steam engine car forced to stop every 10 minutes for its boiling kettle to power up again. Fasten your seat belt because we'll be motoring along by rolling, steaming, crushing, and screaming on our way from here to there!

UP CLOSE

Wheel time is anytime for monocyclist and mechanic Kerry McLean.

Kerry goes for a spin around the racetracks in his hand-built monocycles (also known as monowheels). Founder of McLean Wheel, Kerry has crafted these ultimate machines since the 1970s. He is tinkering with the latest version of the Rocket Roadster, quite possibly the first operational monowheel outfitted with an aluminum block Buick V8 engine with dual quad carbs. After all, he already has one Guinness World Record — why not roll for two?

RECORD 6

Fastest Monowheel

One of the wackiest modes of transport ever devised is the **monowheel**. The monowheel is a one-wheeled vehicle in which the rider sits inside a ring within an outer wheel. There is a hand-controlled brake, but no steering yoke. The rider uses his body to balance the monowheel and lean in the direction he wants to go. The first models were pedal-powered, but modern versions use mechanical engines instead of human energy. On January 10, 2001, Kerry McLean of Walled Lake, Michigan, rolled his way into the record books at 57 miles per hour at Irwindale Speedway in California, USA. Kerry's transportation of choice measures 4 feet in diameter with a former 20.7-cubic-inch snowmobile engine generating 40 horsepower for the **Fastest Monowheel**.

RECORD 7

Earliest Automobile

The automobile shifted horsepower from the "hoof" to under the hood.

Remember the Smethwick Engine of Record 3? Early vehicles relied upon the steam engine to push them down the roads at a top speed of 2.5 miles per hour — that's slower than most people walk! A greater test of patience was the stop-start of the drive. These cars puttered out every 10 minutes before the steam engine cooked up enough power for the next short jaunt.

Nicolas-Joseph Cugnot (1725 - 1804) was a French military engineer whose *fardier a vapeur*, or "steam wagon," converted steam pressure into mechanical energy by turning the wagon's wheels. The **Earliest Automobile** was the first of eventually two military steam tractors. Cugnot completed his revolutionary vehicle at the Paris Arsenal in October 1769. The steam wagon had three iron-rimmed wheels — two in the back, and one in the front supporting a huge steam boiler and engine. The steam wagon was built to haul military equipment for the French Army, with a carrying capacity of 4 tons.

The pneumatic (air-filled) tires we have today were invented in the late 19th century.

Dual Marks

Hybrid cars use a gas engine *and* an electric motor to efficiently produce energy. When the electric motor is running, the resting gas engine saves fuel and reduces smog-producing emissions. The electric motor also assists the gas engine during acceleration.

The Toyota Prius was the first mass-produced hybrid car to be sold globally. Since its launch in 1997, the Prius has sold 250,000 units to become the Best Selling Hybrid on the market. The Prius gets 47 miles per gallon (mpg) in city driving and 60 mpg on the highway.

Auto racing tests both the driver and the vehicle.

The best racing teams take advantage of loopholes whether on the track or in vehicle design. The sport of monster truck racing permits mechanics to relocate the engine from traditionally front-heavy to a balanced mid-body. Officially, truck racing is non-contact. But in reality, bumps and scrapes are bound to happen because the track is barely wide enough to place two of these massive race trucks side by side! Like many other forms of racing, it's the physical contact that makes it such a popular spectator experience.

Torque is the amount of force applied in making gears twist or turn.

RECORD 8
Largest Racing Vehicles

Racing monster trucks became popular during the mid-1980s. Created as outrageous entertainment events during the downtime between car races, these gigantic trucks quickly won fans, and a new sport outgrew the arena. Vehicles in the FIA European Truck Racing Championship weigh at least 12,125 pounds (over 6 tons!) and are powered by 1,000-horsepower turbocharged engines. These monster trucks are the **Largest Racing Vehicles**, exerting an astonishing 2,200 pounds-per-feet of torque in acceleration power. The vehicle's extreme size and weight increase the danger in the event of a crash. The FIA safety rules strictly enforce a top speed of 99 miles per hour, so the drivers put their skills to the test in squeaking past the others to cross the finish line first.

FACT:

Monster trucks entertain all ages by crushing empty cars. Bob Chandler built a fleet of Bigfoot monster trucks (see the cover). The Biggest Monster Truck, Bigfoot 5, is 15 feet 6 inches tall with 10-foot high tires, and weighs 38,000 pounds (not pictured).

BABY STEPS

Even in its infancy, truck racing never included the trailers. The Italian winner of one of the first truck-racing events arrived with a fully loaded trailer, unhooked it, won his race, and then went back to work and completed his delivery.

Although the specialized racing trucks are distant cousins of the highway convoy, speedsters can still get a ticket. In FIA truck races, the penalties range from adding seconds on the finishing time to exclusion from the race. If you're caught going faster than the 99 miles per hour speed limit, then you've lost before crossing the finish line.

Fire fighting began when the first spark was struck.

Early human history of dousing the flames:

- **The first known fire pump was invented in 200 BCE in Alexandria, Egypt.**
- **The first fire department was in ancient Rome.**
- **Firefighters only had two-quart hand syringes to fight the devastating London fire of 1666.**
- **In Colonial America, people left a bucket of water on their stoops in case of a fire at night.**
- **In 1743, Thomas Lote built the first American fire engine.**
- **In the early 20th century, engines replaced horses, and pumps made way for hoses, ladders, and a water tank.**

RECORD 9

Fire Engine with the Greatest Pumping Capacity

A fire **engine** has its own water supply with an engine that pumps water. If the supply runs low, the vehicle can be connected to another water source, such as a fire hydrant. However, a fire **truck**, which is equipped with ladders, tools, fire-fighting equipment, and emergency gear, does not have its own onboard water supply. The most common fire engines, like the ones you may have seen in your own neighborhood, have a diesel pump with a pumping capacity of approximately 2,000 gallons per minute. The **Fire Engine with the Greatest Pumping Capacity** is an 860-horsepower, eight-wheel fire truck manufactured by the Oshkosh Truck Corporation in Wisconsin, USA. It weighs 66.14 tons and produces 49,900 gallons of foam through two turrets in 2 minutes 30 seconds. The Oshkosh fire truck is used for high-intensity aircraft or runway fires (not pictured).

In 2004, there were 1,100,750 firefighters in the USA.

SPOT THE MASCOT

Dalmatians are the traditional firehouse mascot for good reason. This dog breed was highly recognizable with its black-and-white spotted coat. They were trained to run in front of the horse-drawn carts to help clear a path to the fire, a visual alarm to the surprised spectators. The dogs also easily adapted to sharing the stable space with horses, their larger fire-fighting companions.

FACT:

A standard fire truck can fill an Olympic-size swimming pool with 280,000 gallons of water in 2 hours 20 minutes.

UP CLOSE

RECORD 10

Fastest Aircraft with Air-Breathing Engine

NASA's Hyper-X program designed to explore rocket alternatives for space travel also set the Guinness World Record for **Fastest Aircraft with Air-Breathing Engine**. A B-52B bomber plane took off from Edwards Air Force Base in California, USA, on November 16, 2004. The bomber launched the unmanned X-43A research vehicle attached to a *Pegasus* rocket booster. When the aircraft separated from its booster, it accelerated to hyperspeed on scramjet power. Flying 7,000 miles per hour — nearly 10 times the speed of sound at Mach 9.68 — the X-43A reached an altitude of 115,000 feet over the Pacific Ocean. This effort was its second record-setting hypersonic flight, surpassing its previous speed record of Mach 6.8 set in March 2004. This outing marked the third and final flight for the X-43A. It burned out its engine before descending into the Pacific Ocean.

FACT:

Traveling at 7,000 miles per hour would transport you from Earth to the moon in approximately 34 hours.

Combustible energy takes the form of fire.

You can make fire by rubbing two sticks together. This is combustible energy, a chemical process caused when a substance (wood) reacts against molecules (air) to create heat and light (fire). NASA's "scramjet" or air-breathing engine is a revolutionary design free of any moving parts. Compressed air passing through the engine transforms into supersonic combustible energy. This type of engine can be flown like an airplane by controlling its acceleration, unlike a rocket that continually produces a full thrust. Because scramjets don't have to carry heavy oxygen tanks, these planes offer greater navigation capabilities at ultra-high speeds during the first stage of Earth orbit.

Gears and Gadgets

Two hundred years ago, the Industrial Revolution converted our people-powered lives into a mechanized society. Today's Digital Revolution has altered the means and methods of our communication and daily experiences. Modern inventors have transformed science-fiction fantasy into the latest in must-have gadgetry, but the greatest ideas take time to materialize into metallic reality. It took years for our current crop of mobile entertainment devices, from the telephone to the video game, to shrink down into easy-to-carry sizes.

We use more hand-held devices to translate our needs to machines.

An American inventor and pioneer in human-computer interaction, Dr. Douglas C. Engelbart began working at the Stanford Research Institute in the 1960s. There, he and several engineers developed many critical tools used today in our internationally linked civilization. On October 29, 1969, Engelbart's lab electronically linked up to a computer node in the UCLA lab, establishing a computer network that would eventually form the backbone of the Internet.

RECORD 11

Oldest Computer Mouse

Computer guru Dr. Douglas C. Engelbart fundamentally changed the accessibility of computers, from highly specialized machinery restricted for use by scientists, to a user-friendly tool for all ages and backgrounds. He patented his invention in 1964. The **Oldest Computer Mouse** looks like a wooden shell with two metal wheels. The patent application lists an X-Y position indicator for a display system, and the device was to be used with a graphical user interface. It was nicknamed "mouse" because the wire peeking out of the end of the shell looked like a tail. The computer mouse didn't enter common usage until the 1980s with variously shaped options, but this one small tool is indispensable in the everyday interface with computers.

RECORD 12

Earliest Color TV Transmission

There were many scientists on both sides of the Atlantic working on the concept of transmitting electronic images, and there were competing ideas on how best to create this mechanical device. On July 3, 1928, groundbreaking Scottish inventor John Logie Baird successfully demonstrated the **Earliest Color TV Transmission** during a demonstration at his studios in Long Acre, London, UK. John's mechanical television was used by the BBC, the leading television channel in England, from 1929 until 1937.

TEST PATTERN

Three years prior to transmitting color images, John Logie Baird amazed the scientific community with earlier successful experiments using his "televisor." In February 1924, he transmitted a static, brownish image of a ventriloquist's dummy. On October 30, 1925, he sent the first transmission of a moving image — a ventriloquist's dummy head. Eager to get the word out, John went to the newspaper office of the *Daily Express*. The news editor refused to believe that someone had invented a machine capable of showing electronically transmitted pictures.

FACT:

Among the images John Logie Baird transmitted were red and blue scarves, a policeman's helmet, and a bouquet of roses.

Accidents lead to inventions.

On March 10, 1876, Alexander Graham Bell spilled some battery acid and a wire, lying in the liquid, transmitted his voice's vibrations as he called out to his lab assistant, Thomas Watson. "Mr. Watson. Come here please. I want you." These were the first words spoken using a communication device named the telephone. In 1915, Bell and Watson celebrated the connection of the first transcontinental phone lines. From New York City, Bell repeated his famous words over a fully assembled telephone. Watson replied that he needed a week to arrive, because back then he was in San Francisco, California!

RECORD 13

First Portable Cell Phone

The concept for a mobile phone popped up in 1947 at Lucent Technologies' Bell Labs in New Jersey, USA. But it was Martin Cooper of Motorola Corporation who made cellular history on April 3, 1973. While strolling through New York City, he chatted on his hand-held DynaTAC phone with rival developer Joel Engel, Head of Research at Bell Labs. The race to get a mobile phone into the marketplace began. Given just 6 weeks to design their wireless wonder, Martin and seven other co-inventors crossed the finish line first by filing their patent on September 16, 1975. After 10 years and 100 million dollars, Motorola introduced the **First Portable Cell Phone** to consumers. The DynaTAC 8000X became known as the "brick" because it weighed a whopping 2.5 pounds, cost $3,995, and needed a recharge every 30 minutes. Today, more than 180 million Americans use cell phones that weigh less than 3 ounces and fit in the palm of a child's hand.

FACT:

Many inventors were working on the idea of transmitting the human voice. Elisha Gray arrived at the patent office just a few hours behind a man named Bell.

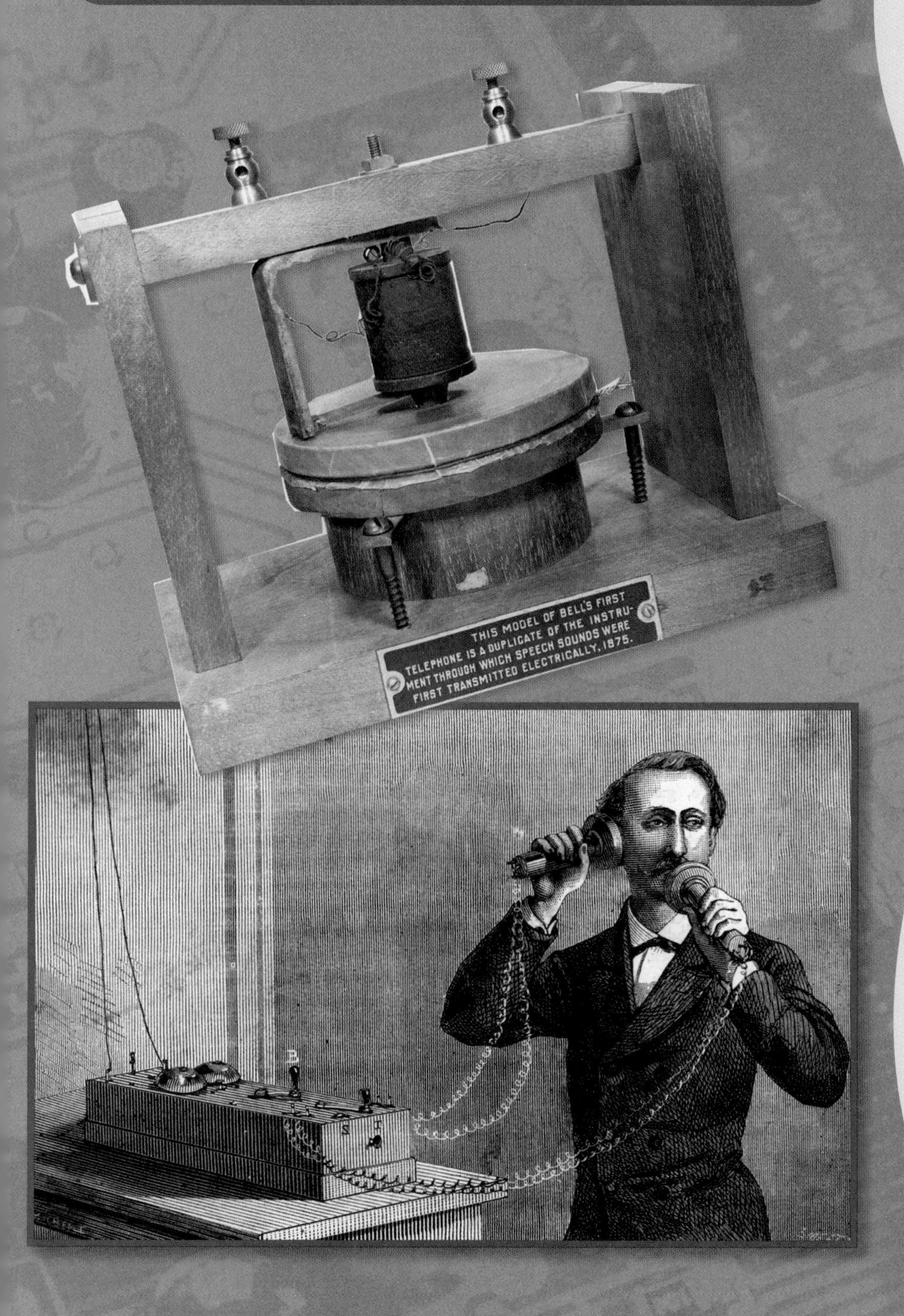

411 ON BELL

Like Leonardo da Vinci, Alexander Graham Bell crafted plans for many inventions: hydrofoils, flight vehicles, metal detectors, and the wireless photophone. His 1880 wireless device transmitted sound on a beam of light and is the grandfather of cordless phones. Because his mother and wife were deaf-mutes, Bell was fascinated with sound and its ability to travel through waves. Bell "spoke" to his mother by pressing his mouth to her forehead and talking in low, resonant tones. The vibrations allowed her to understand his speech. His telephone transmitter designs had elements that emulated the structure of the human ear.

The computer has revolutionized the way we work, talk, and play.

Bites from computer gaming history:

- ***Spacewar!* (1961) created at MIT and was the First Computer Game.**
- **Nolan Bushnell of Nutting Associates designed *Computer Space* (1971), the Earliest Video Arcade Game.**
- **A mustachioed, red-hat plumber named Mario debuted in Nintendo's *Donkey Kong* (1981).**
- **Forty percent of American households own a Nintendo game system.**

The Best-Selling Video Game is Nintendo's *Super Mario Brothers* with 40.23 million copies sold worldwide.

FACT:

The 26 games featuring Mario have sold more than 152 million copies in total since 1983.

RECORD 14

Most Successful Coin Operated Game

Tohru Iwatani of the Japanese company Namco designed the **Most Successful Coin Operated Game**. A pizza missing a single slice was Tohru's inspiration for a game about eating. It was first called *PUCK MAN* because it sounds like the Japanese word *pakupaku* for "he eats, he eats." Seventeen months later, Tohru's clever maze-like game debuted in Japan to a lackluster reception. The name changed for the game's American debut. *PAC-MAN* exploded in popularity when American boys, girls, young, and old became obsessed with steering the yellow mouth through the levels. From its launch in 1981 until 1987, a total of 293,822 *PAC-MAN* arcade machines had been installed worldwide. It is estimated that during its 25 years of electrified life, *PAC-MAN* has been played more than 10 billion times!

PAC-PERFECTION

On July 3, 1999, American Billy Mitchell achieved the First Perfect PAC-MAN Score in a tournament against two Americans and two Canadians, monitored by Twin Galaxies, the authority on computer-gaming high scores. Using just one quarter and "one life," Billy played for almost 6 hours nonstop, eating every blue man, energizer, dot, and fruit on all 256 boards, until he achieved the maximum potential score of 3,333,360 points and the game ran out of memory.

Music lovers used to be stuck at home with a turntable.

In 1963, the Philips Electronics Company invented the cassette tape. Sound was converted to electrical impulses and saved on a magnetic cassette tape. It wasn't until 1979 when two inventors in Japan changed the way the world listened to music. Today we choose where, when, what, and how to enjoy our favorite tunes. Mobile devices from mini-mp3 players to satellite radio subscriptions offer music-lovers an astounding range of options for ordering music on-the-go.

Akio Morita came up with "Walkman" and branded the device with this one name in every country.

RECORD 15

Most Successful Portable Music System

Masaru Ibuka and Akio Morita of Japan's Sony Corporation had a hit on their hands with the 1979 launch of their Sony Walkman TPS-L2. This simple, pocket-sized cassette player was powered by replaceable batteries and featured two headphone jacks for two people to listen to music together. One of the keys to the device's rapid success was Masaru's idea to miniaturize the system's headphones to a featherweight size, weighing only 1.5 ounces! The Sony Walkman became the **Most Successful Portable Music System**, selling between 200 and 250 million total units by April 2001. The word "walkman" now refers to any mobile device that can play music, whether on tape, radio, CD, or digital.

UP CLOSE

FACT:

Different names for the Walkman were originally planned for different countries: *Soundabout* for USA, *Freestyle* for Sweden, and *Stowaway* for Britain.

A SONY REVOLUTION

In 1946 Tokyo, Akio Morita and Masaru Ibuka founded a small electrical repair company to develop new and exciting gadgets. Since then, Sony Corporation has created more life-changing machines:

- **3.5-inch micro floppy disk drive (1981)**
- **MiniDisc – An ultra compact optical disk (1991)**
- **Sony Playstation – videogame system (1995)**
- **WebTV – Internet through TV (1996)**

Mr. & Mrs. Robot-o

Mechanical men and women have migrated from the realm of science-fiction films into affordable helpmates for our homes. Is there a chore you don't like doing? Whip out your remote control and summon your robotic vacuum, lawn mower, or personal humanoid assistant at the press of a button. What task would your ideal robot complete for you?

RECORD 16

Most Emotionally Responsive Robot

It must be Kismet. That's the name of the **Most Emotionally Responsive Robot** created by Cynthia Breazeal at MIT's Artificial Intelligence Lab in Boston, Massachusetts, USA. Kismet can recognize and respond to human emotions by assessing visual and auditory cues in real time. Powered by 15 networked computers and 21 motors, Kismet may only be a robotic head right now, but what a head it is! This robot has advanced perception, motivation, behavior, motor skills, and physical movement capabilities. It can interact in a surprisingly human-like fashion. Some of its numerous facial expressions include winking, smiling, gazing, frowning, raising its eyebrows in surprise, or perking its ears in interest. Kismet will even fall asleep if overstimulated.

Will human-looking robots run our future?

Every holiday, a new robotic toy appears on the shelves claiming to be better than the real thing. How could these machines be as sensitive, caring, responsive, and unpredictable as your flesh-and-blood parent or pet? Flash forward a few years to the workbenches of artificial intelligence experts who are fine-tuning their masterpieces, and you might be surprised to see just how expressive these metallic faces can be. Some people believe the more *human* the robot, the greater chance it will become the caregiver of the future.

Household chores worn you out?

We invent machines to improve the quality of our lives, and that includes the cleanliness of our homes. For those who have an extra $49,508 in pocket change, the TMSUK-4 robot went on the market as the Most Expensive Commercially Available Robot on January 23, 2000. The Japanese Thames Company exhibited their 220-pound, humanoid-shaped, 4-foot tall, personal assistant in Tokyo, Japan (central photo on page 30). TMSUK-4 responds remotely via link-up to your mobile phone. It will run errands, give a massage, or simply listen to you without interruption.

Check out Record 23 to read about how Friendly Robotics teamed up with another record-holder to create more ultimate machines.

UP CLOSE

RECORD 17

Best-Selling Robot Lawn Mower

Stop stressing about the lawn now that the **Best-Selling Robot Lawn Mower** is here! Israel-based Friendly Robotics developed this cutting-edge mechanism to save time and energy. The RL500 Robomower is the replacement model for the classic Robomow. Launched in March 1999 by the Thames, Oxfordshire, UK subsidiary, more than 5,000 units have been sold worldwide since February 2001. Battery-powered, this 3-foot long unit can clip 6,000 square yards of grass in one trip — that's about the size of two tennis courts! But you have to set up lawn boundaries marked by pegs and wires to assist the robot's sensors in mapping out its mow area. The user presses "go" and Robomower zigzags across the lawn. Its safety features stop it from mowing down the family, but it's up to you to protect the flowers.

RECORD 18

Most Advanced Vacuum Cleaner

The stylish yellow and silver Dyson DC06 is the **Most Advanced Vacuum Cleaner**, with 50 sensory devices to scan and map a room in 4 images per second. It shares these images with three onboard computer "brains" to plan the most efficient cleaning path. The self-propelled vacuum weighs 19.84 pounds and can make up to 16 decisions per second as it navigates around rooms, children, and pets! It's also smart enough to recognize and avoid falling down a staircase. Want to know what mood your vacuum is in today? The Dyson DC06 has a color display. Blue means it's pleased with the job, green says it's occupied in moving around an obstacle, and red flashes in warning of a technical problem. Did you leave your socks on the floor again?

FACT:

Ives McGaffey patented a sweeping machine to clean rugs in 1869.

PUFFED OUT

Hubert Cecil Booth in Britain patented a powered vacuum cleaner in 1901. After seeing a device used in trains that blew dust off chairs, he thought it would be more productive to have a machine that sucked up the dust instead. He invented the Puffing Billy and staged a dramatic demonstration. A horse-drawn wagon was parked outside the building to be cleaned. Long hoses were brought through the windows, and a gas-powered engine provided the suction power. Unfortunately, all that huffing and puffing did not blow any money Booth's way, and Puffing Billy was stored in history's closet.

Robots affect our lives every day.

Built on legs or wheels, with jets or wings, robots can show off their skills on sand, mud, and rough ground. Military forces use these models to gather information from dangerous places where humans shouldn't go, such as mine fields. Robots also assist in search-and-rescue missions, and space explorations. Who knows what type of life our non-human friends may find out there? Maybe our robotic explorers will encounter alien robots. We hope their interfaces are compatible for a peaceful data exchange.

RECORD 19

Highest Jumping Robot

Rush Robinett was catching grasshoppers for fishing lures when the proverbial light bulb popped, or hopped, over his head. The random, slightly awkward way grasshoppers jump anywhere and everywhere gave him the inspiration to design small, mobile, jumping robots. Rush is a robot designer at the Intelligent Systems and Robotics Center of Sandia National Laboratories in New Mexico, USA. The **Highest Jumping Robot** is about the size of a grapefruit. A combustion chamber inside a spherical plastic shell activates a piston that propels the robot to heights of up to 30 feet. One jumper can travel 6 feet from its starting point. Fellow designers Barry Spletzer and Garry Fischer devised propane fuel to power the hoppers. A miniature tank holds about ¼ cup of propane fuel — enough for 4,000 hops or 5 miles. If the Moon or Mars landers released a swarm of hoppers to survey a planet, these fueled-up explorers could jump higher and travel farther on lower-gravity planets.

FACT:

At the 2003 Robodex Expo in Japan, a troupe of Monsieur robots performed a ballet, choreographed for them!

RECORD 20

Smallest Robot

The Monsieur I microbot developed in 1992 by Japan's Seiko Epson Corporation is the **Smallest Robot** recognized by the officials at Guinness World Records (see the sidebar for more about the micromechatronic series). In the same year as its debut, Monsieur I won a design award at the International Contest for Hill-Climbing Micromechanisms. Constructed from 97 separate watch parts, the slight robot weighed 0.05 ounces and measured less than 0.06 inches. This light-sensitive mini-bot could fly for about 5 minutes at a speed of 0.4 inches per second. Its designers began tinkering and thinking and, before long, new Monsieurs were flying into the marketplace, but the first Monsieur is the one and only recognized record-holder.

MR. PETITE DEUX

Seiko builds microrobots to showcase its micromechatronic technology. When Monsieur I was recognized as the Smallest Robot, Guinness World Records retired the record because it would be impossible to keep up with the hourly advancements in micro tech. Since then, many engineering companies have developed tinier mechanical gadgets. Seiko's Monsieur II-P is activated by an ultra-thin, ultrasonic motor and can really fly. The super small robot is radio-controlled by wireless Bluetooth and moves at speeds up to 2.7 inches per second — that's about 100 miles per hour! At less than ¾ inch in size, and weighing in at ½ an ounce, you would not even see it speed by.

Best Of . . .

The most outrageous, can't-ever-be-done ideas have generated our most life-changing, can't-live-without machines. Who would have believed that riding up, down, and around steel curves in a metal box is a popular entertainment? Competitive engineering challenges exist to create moneymaking devices for catching pesky rodents or combining water and wind for personal hygiene. We'll look under the cushion of some motorized furniture and take a peek under our own hoods to learn about the biological computer living inside all of us!

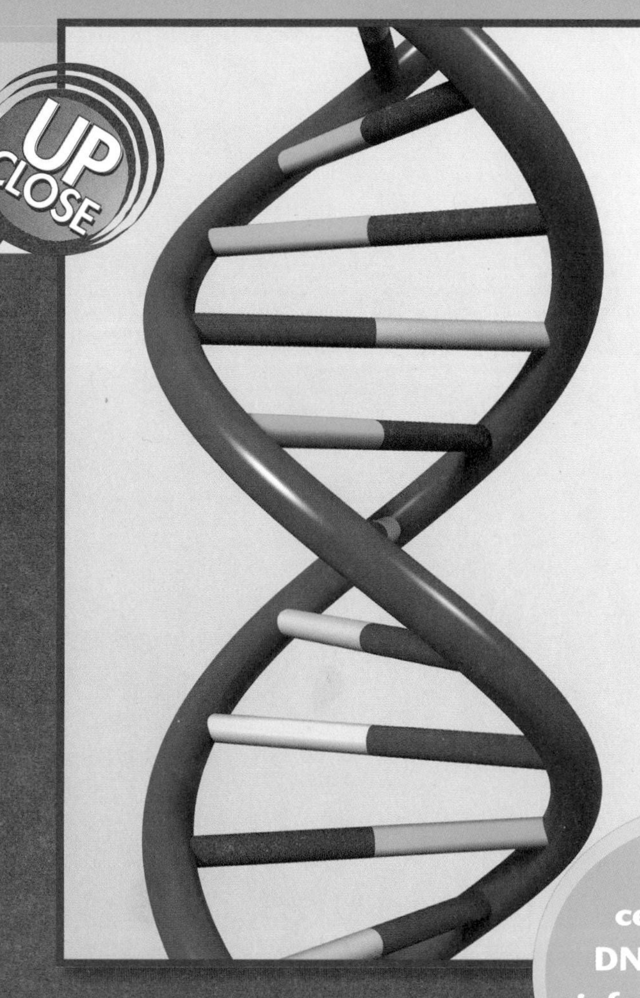

One cubic centimeter of DNA holds more information than a trillion music CDs.

RECORD 21

Smallest Biological Computing Device

Ehud Shapiro and a team of scientists announced the successful design of the **Smallest Biological Computer Device** at Israel's Weizmann Institute of Science on February 24, 2003. The programmable molecular computing machine is actually composed of an enzyme and two DNA molecules, not silicon microchips. The trio's molecular construction and interaction resembles a regular computer's input, software, hardware, and power supply. The DNA provides the input data and the power to complete the tasks! To the naked eye, the DNA computer looks like water in a test tube. But a spoonful of this "computer soup" solution containing 3 trillion devices could perform 66 billion operations per second!

The ultimate computer is already at work preparing for the future — *inside* of us.

Hybrid computers are on the horizon. These machines would use silicon for normal processing and organic DNA co-processors for specific tasks. DNA computing devices could be particularly useful in the biomedical field. DNA computers might monitor an individual's health. Like a "doctor in a cell," it would analyze problems, determine the appropriate prescription, then administer the medicine.

Rollercoasters make people scream, laugh, and cry — all at the same time!

Winter lasts for eight months in St. Petersburg, Russia. In the mid-1600s, a popular outdoor activity was riding wooden ice slides 80 feet high and hundreds of feet long. By the 1800s, the French built thrill rides of wheeled cars locked onto tracks. In 1884, Marcus Thompson designed a coaster similar to the two parallel hills of the Russian ice slides (see sidebar). Today, advances in engineering and design have made these scream machines bigger, faster, and enjoyable all year round!

RECORD 22

Fastest and Tallest Rollercoaster

A screaming thrill ride named Kingda Ka opened on May 20, 2005, at Six Flags Great Adventure amusement park near Jackson, New Jersey, USA. Eager passengers shrieked through the dual record-setting slopes and curves of the **Fastest and Tallest Rollercoaster** during a ride lasting 50.6 seconds. Within 3.5 seconds, a hydraulic launch propels the coaster to its top speed of 128 miles per hour. Riders are safely strapped into their seats as they crest the coaster's tower at 456 feet above the ground — that's as tall as a 45-storied building. Kingda Ka takes a vertical plunge into a 270-degree spiral, giving riders the temporary and exhilarating feeling of weightlessness, before swooping and climbing and gliding its way back into the station.

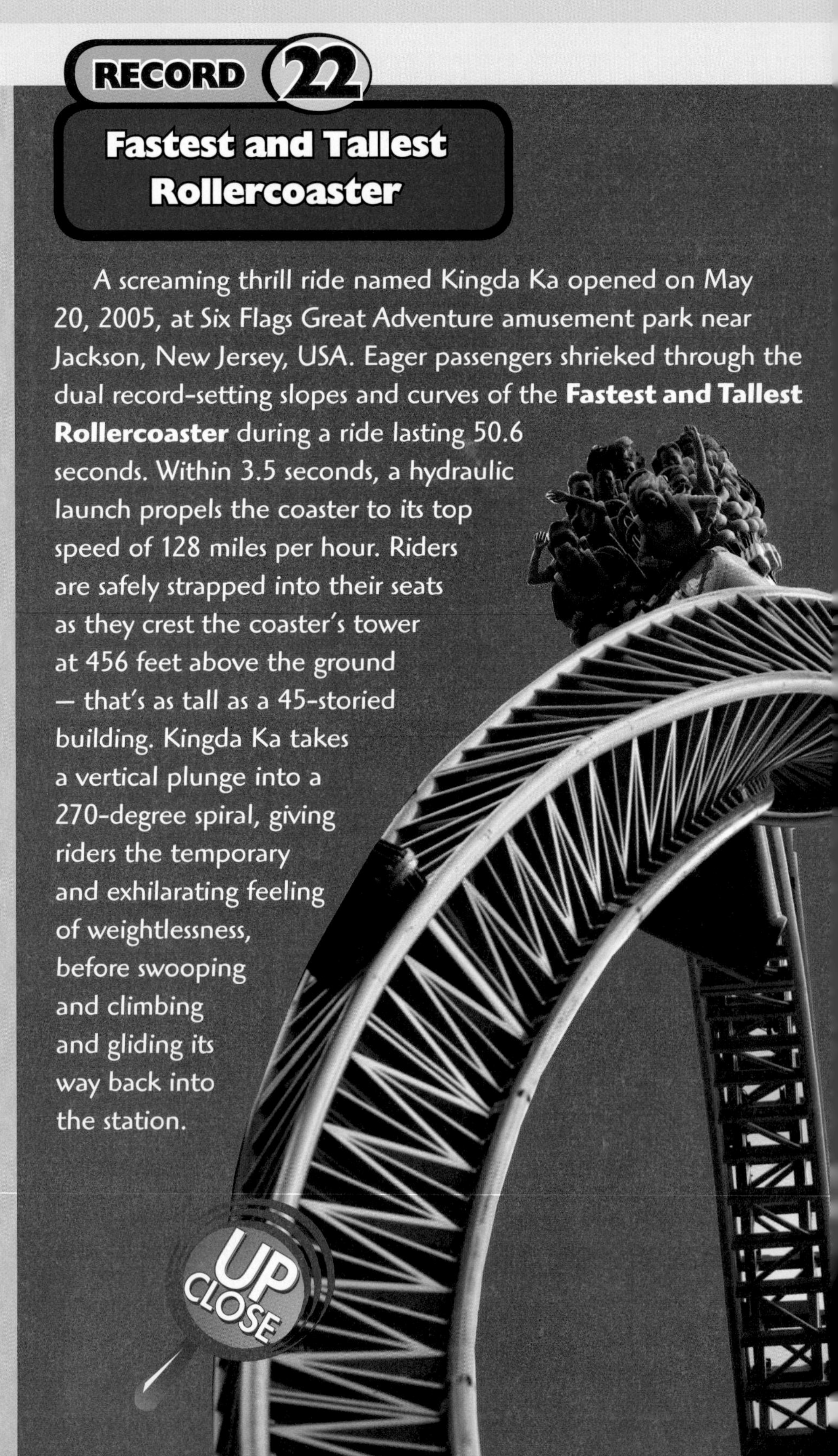

COASTING ALONG

Marcus Thompson discovered that switching between tracks helped the coaster cars handle twists and turns. In 1884, his Switchback Railway made ground-shaking history at Coney Island, New York, USA. Nicknamed Mr. Gravity, Marcus earned hundreds of shrieks and dollars every day from passengers aboard his scream machine. Back then the ride cost a nickel, so there were lots of thrilled people!

In the 1950s, the first tubular steel coaster, the Matterhorn, debuted at Disneyland. The hollow steel allowed the lighter coaster to fly through loops, corkscrews, and inversions that people love to ride.

A **lift hill** is when a chain or cable pulls the cars upward before that first steep plunge.

FACT:

Bakken is the Oldest Operating Amusement Park, entertaining kids of all ages since 1583 in Klampenborg, Denmark.

Because life is all about the journey, you should always travel in style.

Edd China and David Davenport are the creators of real road-hugging furniture. They have been revamping household standards, from bathroom suites to four-post beds, for promotional, fun, and legal UK road driving since March 1999. Their company, Cummfy Banana, is based in Odiham, Hampshire, UK, and continues to offer uniquely customized vehicles for their adventuresome clientele (see sidebar).

RECORD 23

Fastest Furniture

Since the right piece of furniture can improve your home, why not add a set of wheels and improve your road trips? The **Fastest Furniture** is the Casual Lofa, a leopard-print, motorized sofa with a top speed of 87 miles per hour, designed and built by Edd China and David Davenport. Legal to drive on UK roads, the vehicle has traveled 6,219 miles since assembly. Originally a three-wheeler with a Mini 850 cc engine and a Reliant Rialto base, the Casual Lofa was revamped with a Mini 1300 cc engine and four wheels. Several features have changed during the overhaul, but the ride is always enjoyable because a working television entertains two passengers while the driver avoids less-comfortable traffic.

CUMMFY RIDES

Interested in revving up your furniture? Here's a showroom sample at Cummfy Banana:

- **Street Sleeper:** This bed fits 3 and is a dream at 69 miles per hour.
- **Bog Standard:** A motorcycle with its sidecar converted into a shower and toilet. Hydrant hook-up for running water required.
- **Hot-Desk:** Workers on-the-go hit the books on a desk equipped with phone, computer, Internet, fax, and water cooler while this functional mobile office hits 70 miles per hour.

FACT:

Cummfy Banana also built two Robomow cars, commissioned by Friendly Robotics, to promote the Best-Selling Robot Lawn Mower (see Record 17).

The indoor plumbing we enjoy in our homes is a fairly recent invention of modern life.

Today, our most pressing concern is not having enough toilet paper in the bathroom. Imagine not having sewage pipes to flush. What did people do before plumbing? People would use holes in the floor with drains depositing human waste in the rivers.

The White House has 35 bathrooms.

RECORD 24

Most Sophisticated Toilet

A toilet found in India is more than 4,000 years old. Forget privacy, pipes, or paper. Fast forward to today's sophisticated engineering. The Washlet Zoe will flush away your worries. The **Most Sophisticated Toilet** has revolutionized the restroom and broken records with its seven flush functions controlled by remote. The lid automatically opens and closes, and the seat itself is comfortably heated. An audio function creates white noise to drown out any embarrassing sounds. A wand squirts warm water for cleansing, and a warm air jet dries the area. No toilet paper required! Plus, the Washlet Zoe scents the air with an automatic freshener, just in case. The Japanese company TOTO introduced the Washlet Zoe in May 1997. Now there are over 17 million Washlets in use around the world. That's a lot of pampered bottoms!

FACT:

In 1829, the Tremont Hotel in Boston was the first hotel with indoor plumbing. It had 8 toilets on the ground floor.

TALES FROM THE CHAMBER POT

The Ancient Romans constructed a plumbing system with underground sewers. The knowledge of this invention was lost in the Middle Ages, leaving people to use chamber pots or *potties*, and to dump the contents out of windows into the streets below. The streets had drains that led to a water source. This was highly unsanitary and caused widespread disease. Toilets similar to our modern ones became available, but only to the wealthy. Modern sewer systems have made it possible for most homes in the USA to have at least one toilet with pipes leading to a proper sewage-treatment center.

RECORD 25

Biggest Mousetrap

To celebrate its 65th anniversary in 2003, the American pest control company Truly Nolen super-sized John Mast's device and built the **Biggest Mousetrap**. The trap measured 11 feet 7 inches long x 5 feet 6 inches wide and weighed over 600 pounds. The springs were made from 40-foot long steel rods. The trap's snapping rod was powerful enough to catch and crush a small car. On November 5, 2003, Truly Nolen officially sprang their record-setting trap in Miami Gardens, Florida. No cars or mice were harmed during the demonstration.

FACT:

Every year, 400 people apply for mousetrap patents. In 1899, a Pennsylvanian named John Mast answered the mousetrap challenge with the simple snap-trap style seen in stores and on television today. Many believe that Mast's mouse-catching machine is already the best mousetrap.

Building the better mousetrap is every inventor's challenge.

The phrase "to build the better mousetrap" refers to improving upon the previous design of a machine. Approximately 5,000 patents for bait-and-catch rodent devices have been issued in the United States, although only 20 mousetrap designs are financially successful.

ZOOM OUT!

Although our book ends here, your exploration of these ultimate machines and their record-setting stories can continue among the online archives (www.guinnessworldrecords.com) and within the pages of *Guinness World Records*.

Go up close and get involved — it's your world!

Photo Credits

The publisher would like to thank the following for their kind permission to use their photographs in this book:

Cover, title page (main) Bigfoot Car Crushing © 2006 BIGFOOT® 4x4, Inc., (inset) Bigfoot Driver © David Huntoon/Action Images; 3 Monster Truck Head-On © Mark S. Wexler/Getty Images; 4 Grouse © Peggi Miller/iStockphoto; 5 (left), 10 Sam Whittington and *Varna Diablo* © John Cassidy; 5 (center) Montezuma Wind Farm © Charlie Riedel/AP Wide World Photos, (right) Draisienne © The Art Archive/ Musée Carnavalet Paris/Dagli Orti; 6 Grand Coulee Dam Panorama © Michael T. Sedam/CORBIS; 7 (top) Waterwheel © Topham/The Image Works; 7 (bottom) Grand Coulee Dam, 23 John Logie Baird, 25 Alexander Graham Bell and Telephone © Bettmann/CORBIS; 8 Altamont Wind Turbines © Lonny Shavelson/Newscom; 9 Smethwick Engine © David Rowan/Thinktank, Birmingham Science Museum; 11 *Decavitator* © 1992 Steve Finberg; 12 (left) Toyota Prius © Robert E. Klein/AP Wide World Photos, (center) Motoring Monowheel Courtesy of Kerry McLean, (right) X-43a Aircraft © Dryden Flight Research Center/NASA; 13 (left) McLean Rolling © Bill Pugliano/Liaison/Getty Images, (right) McLean Parked © Dan Callister/Liaison/Getty Images; 14, 15 Cugnot's Steam Engine © CORBIS; 15 (top) Toyota Prius © Jeff Christensen/Reuters; 16 FIA Truck Racing © Thomas Frey/Imago/ICON SMI; 17 Monster Patrol Truck © Chuck Burton/AP Wide World Photos; 18 Firefighter © Ted Horowitz/CORBIS; 19 (top) Fire Engine Courtesy of the Oshkosh Truck Corporation, (bottom) Dalmatian Fire Dog © Amy Walters/ iStockphoto; 20 X-43 Vehicle Concept © Marshall Space Flight Center/NASA; 21 (left) Text Messaging © Ranald Mackechnie/Photonica/Getty Images, (center) Playing Ms. Pac-Man © Richard Drew/AP Wide World Photos, (right) Computer © iStockphoto; 22 Computer Mouse Courtesy of SRI International, Menlo Park, CA; 24 Dr. Martin Cooper © Ted Soqui/CORBIS; 26 (top) Nintendo DS © Reuters/Fred Prouser, (bottom) Super-Mario Model © Yoshikazu Tsuno/AFP/Getty Images; 27 (bottom) Ms. Pac-Man Packaging © Ray Tang/REX USA; 27 (top) Pac-Man Game Display, 28 Akio Morita © Neal Ulevich/AP Wide World Photos; 29 (top) Walkman © Ted Thai/Time & Life Pictures/Getty Images, (center) Disk © Morguefile.com, (bottom left) Music Fan © David De Lossy/Photodisc/Getty Images, (bottom right) Playstation © Zachariasen Christian/Corbis Sygma; 30 (center) TMSUK-4 © Toshifumi Kitamura/AP Wide World Photos, (right) Bluetooth Robot © Shizuo Kambayashi/AP Wide World Photos; 30 (left), 32 (left) Robomower Courtesy Grassland Services Ltd., UK; 31 Kismet © George Steinmetz/CORBIS; 32 (right) Robomower Up Close Courtesy of Robo Direct, Inc.; 33 Vacuum Cleaner Courtesy of Dyson; 34 Jumping Robot Courtesy of Sandia National Laboratories; 35 (top) Flying Monsieur Robot © Katsumi Kasahara/AP Wide World Photos, (bottom) Monsieur Robot Up Close Courtesy of Epson; 36 (left) Chamber Pot © Creative Commons, (right) Casual Lofa © David Davenport 1999/cummfybanana.co.uk; 36 (center), 39 Kingda Ka © Stan Honda/AFP/Getty Images; 37 DNA Strand © Ben Greer/iStockphoto; 38 Kingda Ka's Riders © Noah K. Murray/Star Ledger/CORBIS; 40, 41 Casual Lofa © 2006 Ranald Mackechnie/ Guinness World Records; 42 (left) White House © Susan Marton/Morguefile.com, (center and right) Washlet Zoe Courtesy of TOTO USA, Inc.; 43 (top) Tremont Hotel © The Bostonian Society, (bottom) 1823 Lithograph Paris: *"Is It Raining?"* © The Granger Collection, New York; 44 Biggest Mousetrap Courtesy of Truly Nolen of America, Inc.; 45 Monster Truck Side View © Bloomsburg Press Enterprise, Jimmy May/AP Wide World Photos.

Message from the Keeper of the Records:

Record-breakers are the ultimate in one way or another — the youngest, the oldest, the tallest, the smallest. So how do you get to be a record-breaker? Follow these important steps:

1. Before you attempt your record, check with us to make sure your record is suitable and safe. Get your parents' permission. Next, contact one of our officials by using the record application form at www.guinnessworldrecords.com.

2. Tell us about your idea. Give us as much information as you can, including what the record is, when you want to attempt it, where you'll be doing it, and other relevant information.
   a) We will tell you if a record already exists, what safety guidelines you must follow during your attempt to break that record, and what evidence we need as proof that you completed your attempt.
   b) If your idea is a brand-new record nobody has set yet, we need to make sure it meets our requirements. If it does, then we'll write official rules and safety guidelines specific to that record idea and make sure all attempts are made in the same way.

3. Whether it is a new or existing record, we will send you the guidelines for your selected record. Once you receive these, you can make your attempt at any time. You do not need a Guinness World Record official at your attempt. But you do need to gather evidence. Find out more about the kind of evidence we need to see by visiting our website.

4. Think you've already set or broken a record? Put all of your evidence as specified by the guidelines in an envelope and mail it to us at Guinness World Records.

5. Our officials will investigate your claim fully — a process that can take up to a few weeks, depending on the number of claims we've received, and how complex your record is.

6. If you're successful, you will receive an official certificate that says you are now a Guinness World Record-holder!

Need more info? Check out the Kids' Zone on www.guinnessworldrecords.com for lots more hints, tips, and top record ideas that you can try at home or at school. Good luck!